Lisho Mundia

Development of a Geographical Information System - Based Support Tool

AF141246

Lisho Mundia

Development of a Geographical Information System - Based Support Tool

for Integrated Water Resources Management in Zambezi Catchment Area within Zambezi Region, Namibia

LAP LAMBERT Academic Publishing

Impressum / Imprint

Bibliografische Information der Deutschen Nationalbibliothek: Die Deutsche Nationalbibliothek verzeichnet diese Publikation in der Deutschen Nationalbibliografie; detaillierte bibliografische Daten sind im Internet über http://dnb.d-nb.de abrufbar.
Alle in diesem Buch genannten Marken und Produktnamen unterliegen warenzeichen-, marken- oder patentrechtlichem Schutz bzw. sind Warenzeichen oder eingetragene Warenzeichen der jeweiligen Inhaber. Die Wiedergabe von Marken, Produktnamen, Gebrauchsnamen, Handelsnamen, Warenbezeichnungen u.s.w. in diesem Werk berechtigt auch ohne besondere Kennzeichnung nicht zu der Annahme, dass solche Namen im Sinne der Warenzeichen- und Markenschutzgesetzgebung als frei zu betrachten wären und daher von jedermann benutzt werden dürften.

Bibliographic information published by the Deutsche Nationalbibliothek: The Deutsche Nationalbibliothek lists this publication in the Deutsche Nationalbibliografie; detailed bibliographic data are available in the Internet at http://dnb.d-nb.de.
Any brand names and product names mentioned in this book are subject to trademark, brand or patent protection and are trademarks or registered trademarks of their respective holders. The use of brand names, product names, common names, trade names, product descriptions etc. even without a particular marking in this work is in no way to be construed to mean that such names may be regarded as unrestricted in respect of trademark and brand protection legislation and could thus be used by anyone.

Coverbild / Cover image: www.ingimage.com

Verlag / Publisher:
LAP LAMBERT Academic Publishing
ist ein Imprint der / is a trademark of
OmniScriptum GmbH & Co. KG
Heinrich-Böcking-Str. 6-8, 66121 Saarbrücken, Deutschland / Germany
Email: info@lap-publishing.com

Herstellung: siehe letzte Seite /
Printed at: see last page
ISBN: 978-3-659-79602-9

Zugl. / Approved by: Namibia Geographical Information Technologies Cc

Dedication

We dedicate this work to United States Agency for International Development (USAID); National Aeronautics and Space Administration (NASA); and the Regional Centre for Mapping of Resources for Development (RCMRD) for their trust in the Namibia Geographical Information Technologies cc, (NGIT) company.

About NGIT, RCMRD and Servir Africa

The project is the initiative of the Namibia Geographical Information Technologies cc, (NGIT) a company registered with the Ministry of Industrialisation, Trade and SME Development in Namibia to provide and promote innovative GIS, surveying & related spatial technologies to its clients and business associates. NGIT is committed to:

- rendering innovative GIS solutions and promoting new spatial technologies;
- ensuring to match and customise the data to meet the specific requirements of our clients and business associates;
- helping promote educational institutions;
- assisting professionals and businesses to use and apply geo-information technology to better manage their area of responsibility; and
- assisting clients in successfully deploying, recognising, applying and administering GIS, surveying and database management.

The project is funded by Servir, a regional visualisation and monitoring system that ingrates Earth Observations such as satellite imagery and forecast models, with situ data and other information for timely decision-making. SERVIR is a joint development initiative of the United States Agency for International Development (USAID) and the National Aeronautics and Space Administration (NASA). SERVIR-EASTERN AND SOUTHERN AFRICA is implemented in partnership with Regional Centre for Mapping of Resources for Development (RCMRD) and works towards establishing itself as a regional resource centre in Eastern and Southern Africa by developing relevant geospatial applications, and increasing access to data and decision-support tools on different thematic areas.

ii

List of Abbreviations and Acronyms

FGD	Focus Group Discussions
GIS	Geographic Information System
GWP	Global Water Partnership
IPCC	Inter-governmental Panel on Climate Change
IWRM	Integrated Water Resource Management
KMTC	Katima Mulilo Town Council
MAWF	Ministry of Agriculture, Water and Forestry
MDG	Millennium Development Goals
MET	Ministry of Environment and Tourism
MLR	Ministry of Lands and Resettlements
MRLGHRD	Ministry of Regional, Local Government, Housing and Rural Development
NASA	National Aeronautics and Space Administration
NGIT	Namibia Geographical Information Technologies
NGOs	Non-Governmental Organizations
NSA	Namibia Statistics Agency
RCMRD	Regional Centre for Mapping of Resources for Development
SWOT	Strengths, Weaknesses, Opportunities and Threat
UN	United Nations
USAID	United States Agency for International Development

Disclaimer

This report was produced in the framework of the project *'Development of a Geographical Information System - Based Support Tool for Integrated Water Resources Management in Zambezi Catchment Area within Zambezi Region, Namibia'*. The project funded by Servir, a regional visualisation and monitoring system that ingrates Earth Observations such as satellite imagery and forecast models, with situ data and other information for timely decision-making. SERVIR is a joint development initiative of United States Agency for International Development (USAID) and National Aeronautics and Space Administration (NASA). SERVIR-EASTERN AND SOUTHERN AFRICA is implemented in partnership with the Regional Centre for Mapping of Resources for Development (RCMRD).

If you have any concerns or require clarification on any issue contained in this document please feel free to contact Namibia Geographical Information Technologies cc.

Executive Summary

This report is deliverable of the project '*Development of a Geographical Information System - Based Support Tool for Integrated Water Resources Management in Zambezi Catchment Area within Zambezi Region, Namibia*'.

Consultative meetings, Focus Group Discussions (FGD) and workshops on ancillary data collection were held between 26 August and 10 October, 2014, at the Katima Mulilo National Youth Resource Centre and at various parts of the eight constituencies of the Zambezi region. Training on GIS and the interactive GIS - based digital atlas was held from the 12th to the 14th November 2014. The participants were drawn from Zambezi Regional Council (ZRC); Ministry of Agriculture, Water and Forestry (MAWF); Ministry of Education; Ministry of Lands and Resettlements (MLR); Katima Mulilo Town Council (KMTC); Namibia Water Cooperation Ltd and the Namibia Statistics Agency (NSA).

The purpose of the consultative meetings, FGD and workshops were to gather information on farming practices, the use of GIS, the impacts of flood and climate change, mitigation and adaptation measures in the context of Integrated Water Resources Management (IWRM) in the Zambezi region. The participants shared past and present efforts in introducing GIS - based support systems for planning in the region. The participants cherished the workshop as it discussed some of the Strengths, Weaknesses, Opportunities and Threats (SWOT) analysis of the region.

The consultative meetings, FGD, workshop and trainings resulted in an increased understanding of various aspects such as the need for GIS – based support tool to support the IWRM in the Zambezi region. The participants emphasised the need for training on GIS and GIS – based support tool, the need for continuous consultation, engagements in the developments of other methods for IWRM to be a reality and forming of a technical team for IWRM in the Zambezi region.

The concepts and results of the consultative meetings, FGD, workshop and training are presented in this report.

Table of Contents

List of Figures

List of Tables

Acknowledgement

We are grateful for the support of the Zambezi Regional Council (ZRC); Ministry of Agriculture, Water and Forestry (MAWF); Ministry of Lands and Resettlements; Ministry of Education; Katima Mulilo Town Council (KMTC); Namibia Water Cooperation Ltd; Namibia Statistics Agency; Polytechnic of Namibia and University of Namibia, for attending the consultative meetings, workshops, training, sharing their knowledge and experiences. We also want to thank the above stated organisations and ministries for providing us with the necessary ancillary data.

In addition, we want to acknowledge the participation and contributions of commercial farmers and subsistence farmers in various consultative meetings in response to our questions concerning flood impacts, mitigation and climate change adaptation measures in the Zambezi region.

1. Project Narrative

1.1 Project Description

The project is titled: *Development of a Geographical Information System - Based Support Tool for Integrated Water Resources Management in Zambezi Catchment Area within Zambezi Region, Namibia.*

Given the important role of information in IWRM, a prerequisite for supporting this is the provision of basic data collected over space and time, that allows an understanding of the environmental, social, cultural and economic dynamics of an area to be developed (McDonnell, 2008). Participatory techniques such as Focus Group Discussions (FGD) are extremely useful to gain deeper insight into particular issues affecting a community. However, since the information to be generated are qualitative in nature it is harder to report in a standardised manner or compare with information from other locations in anything more than a descriptive manner (United Nations [UN] - Habitat, 2006:118). It is important in determining problems and areas through holistic participatory approaches such as a FGDs, GIS - based tool for gathering and incorporation of knowledge at the local level than for national reporting of progress only. In support, Kjelds, et al. (2005:513) emphasised that "good water management constitutes a major challenge and calls for integrated and holistic planning. Decision made now affects all and will have great impacts on generations to come."

Many scientific literatures, such as the Dublin Principles of (1992), Kluge et al. (2006), Klintenberg et al. (2007), Inter-governmental Panel on Climate Change [IPCC] (2008), Global Water Partnership [GWP] (2008) & Biggs et. al (2008) stated that there are missing elements at the technical level in IWRM. These elements include relevant datasets, rational charging policies, water conservation, water management, and the use of computer models in decision-making.

Water supply and availability has been identified as a main problem of Integrated Water Resources Management (IWRM) implementation in other parts of the Zambezi region. Within the IWRM framework, the Geographical Information System (GIS) - based tool was developed and adapted in order to support spatial decision-making related to water supply, availability and accessibility within the Zambezi basin of

1

Zambezi region. The tool comprises GIS, particularly a comprehensive digital atlas as an outcome of the research process. The results was generalised concerning their consequences beyond the Namibian case.

1.2 Goal and Objectives

The main goal of the study was to develop the GIS - based support system for IWRM in Zambezi catchment area within the Zambezi Region, Namibia. The objectives were investigated in the context of "Improvement of Water Supply and Availability" focussing on the Zambezi basin in Namibia under the theoretical framework of IWRM. The main objectives of the research were to:

• develop a GIS - based tool to support planning and implementation of IWRM in Zambezi region using a participatory approach.

• evaluate the functions and capabilities of the tool using a Strengths, Weaknesses, Opportunities and Threats (SWOT) analysis technique through participatory approach.

• incorporate the produced maps, graphs, tables, pictures, legislations and results of participatory approaches into the GIS - based tool for decision makers to support the improvement of water supply and availability.

• incorporate climate change issues throughout the project starting with the research and data collection phase to address actual vs. perceived climate impacts; particularly as it relates to population density, flood impacts, agricultural threats, and soil degradation.

1.3 Hypotheses of the Study

These research objectives were linked to the following hypotheses which were addressed during the research:

• The usage of GIS, RS and stakeholders' engagement within IWRM processes can help to improve information awareness, capacity building and access to spatial information.

• The study contributes to a better spatial understanding of IWRM processes and a better decision-making related to water supply, availability and accessibility.

• Subjective perceptions of stakeholders are relevant for decision making and differ from factual knowledge approved by scientific methods.

1.4 Background and Problem Statement

There was no developed GIS - based tool in IWRM for Zambezi basin in Namibia. Water resource spatial data are scattered in various organisations, and no analysis was performed to make the data readable for non-GIS experts. This comes from poor planning and unclear understanding of water resources management by decision makers. Kluge et al. (2006) & GWP (2008) stated that "a holistic approach is required in spatial dimension, decision support system or GIS."

Spatial data gathered and generated by this project were used as input to the data required for development of a GIS - based tool to guide decision makers on how to harmoniously use and conserve water and resources. There has never been a greater need for a GIS - based tool to help in evaluating the applicability of complex environmental models for the Zambezi basin (Karnatak, Saran, Bhatia, & Roy, 2007).The main factors, which drive people to manage water resources, were identified. GIS mapping, visualisation, modelling, maintaining, spatial analysis capabilities and spatial data representation were important components of the GIS - based tool. The dynamic interactive digital atlas environment was the platform of choice for their deployment.

The developed GIS - based tool and the applied participatory approaches of this project were monitored and evaluated to determine their sustainability in supporting IWRM process in Namibia. GWP (2008:92) also confirms that "monitoring systems need to generate information showing the degree and extent to which basin management plans, strategies and programmes are changing the state of water resources, socio-economic and ecological conditions in the basin."

1.5 Project Stakeholders

The project used a diverse number of stakeholders where data, perceptions, opinions, facts and knowledge were gathered and shared. Engaging stakeholders involves establishing good channels of communication between the researcher and various stakeholders. The process then maintains a constructive relationship between the two parties. Through this relationship, stakeholders can have their say and the researcher can listen and respond. Table 1 shows the stakeholders engaged in this project.

Table 1: Engaged stakeholders in this project

Stakeholders	Specific names
Government Ministries	Ministry of Agriculture, Water and Forestry; Ministry of Environment and Tourism; Ministry of Lands and Resettlements; Ministry of Regional, Local Government, Housing and Rural Development i.e. Zambezi Regional Council (ZRC); and Ministry of Education.
State-Owned Enterprises	Katima Mulilo Town Council (KMTC); Namibia Water Cooperation Ltd; Namibia Statistics Agency; Polytechnic of Namibia; and University of Namibia
Private organisations	Namibia Geographical Information Technologies Integrated Rural Development and Nature Conservation (IRDNC)
Local communities	Commercial farmers Subsistence farmers Other local communities

2. Literature Framework

2.1 Concepts of IWRM

Biggs et al., (2008:01) stated that "IWRM is a relatively new concept formally introduced by the international professional water management community with the Dublin Principles for water in1992." This research was relevant in Namibia and contributed not only to the development and implementation of IWRM spatial support tools in Zambezi region, but also to the Millennium Development Goals (MDG) in Namibia ensuring, in particular environmental sustainability as stated in goal number 7, and to halve the proportion of people without sustainable access to safe drinking water.

According to McDonnell (2008:132) "ideas for linking our understanding of engineering and the natural science of water to the social, cultural and political context of an area have been muted for over 70 years, but the notion of IWRM became firmly entrenched in discussions on policy and water use during the last 15 years." The need to integrate has gained increasing credence as the interconnectedness of the many domains of water resources management was appreciated (Braga, 2001; Jonch-Clausen & Fugl, 2001; cited in McDonnell (2008:132).

Interactions and feedback from the natural or human environments have compromised water management projects in many areas of the world. McDonnell (2008:132) further stated that "In more recent years this has been expanded to include other dimensions and leading proponents such as the GWP (2008) perceive it as a new water governance and management paradigm which if effective, could give long-term solutions to water problems. This is advocated through a move away from top-down, supply-led solutions dominated by the adoption of technology, towards a more decentralised basis with a consideration of water in its larger, more holistic context and an appreciation of local ideas and demand management." This concept is of course welcomed and embraces the principles adopted by various governments in Dublin in 1992, as stated already in this report.

2.2 GIS for IWRM

The term 'Geographical Information System' describes an information system. In this project, GIS was used as a computer system with emphasis on spatial data and functions to be able to generate information that can lead to efficient decision-making in spatial planning. The definitions of GIS are important in this project to provide an understanding as to how GIS was involved. Clarke (1997:8) defines GIS as "a powerful set of tools for storing and retrieving at will, transforming and displaying spatial data from the real world for a particular set of purposes." Chrisman (1997, cited in Clarke, 1997:17) however, defined GIS as "organised activity by which people measure and represent geographical phenomena, and then transform these representations into other forms while interacting with social structures."

Recently, GIS has been applied to diverse fields to assist experts in analysing various types of spatial data and dealing with complex situations. These fields include business, government, education, tourism, transportation, and utilities or natural resources management. GIS plays an essential role in helping people collect data, analyse the related spatial data as well as displaying data in different formats. The GIS capabilities have increased and are being extended to include more applications, in specific market sectors.

McDonnell (2008:140) stated that "information systems are needed to manage and share data, and their design and structure should be developed following extensive discussions with potential data providers and users to ensure it will meet future requirements, as well as present requirements. Discussions will also help stimulate a communal sense of ownership which will help to maintain the system after the initial establishment stages are completed."

Software implementation of the GIS – based support tool included the collection of necessary input datasets (RS, GIS-layers, statistics, tables, photo, etc.) and validation. Conceptualise, setup and implement user-friendly georeferenced digital atlas using Geopublisher software. The current vector and raster data, updated by different ministries in Namibia were used. The production of a digital atlas helped improve the integration of a variety of spatial information sources such as GIS and remote sensing (i.e. thematic maps) as well as results of participatory approaches to help support the IWRM.

2.3 The Value of Maps in IWRM

Maps are essential tools for water management organisations to communicate their plans to other parties. They need to communicate internally with members and externally to other experts, water users, government, partners, donors, and investors. Using maps leads to a greater shared understanding of important issues about natural resource management and land use planning (Tagg & Taylor., 2006).

A map is usually recognised as being a simplified, generalised and reduced representation of a part of the curved earth on a flat sheet of paper. Wade and Sommer (2006:130) defined a map as a "graphic representation of the spatial relationships of entities within an area." According to Robinson, Morrison, Muehrcke, Kimerling and Guptill (1995:4-5) "a map has two important functions:
- It serves as a storage medium for information which humanity needs.
- It provides a picture of the world to help us understand the spatial patterns, relationships, and complexity of the environment in which we live."

Robinson et al. (1995:09) defines a map as a graphic representation of the geographical setting. Cartography is more than a map. It is the making and study of maps in all aspects (Robinson et al., 1995:09). Wade and Sommer (2006:27) in agreement with Robinson et al. (1995) defined cartography as "the art and science of expressing graphically, usually through maps, the natural and social features of the earth." Natural and cultural phenomena are represented on the map by unique symbols so that they are easily identifiable. Clarke (1997:8) stressed that understanding the way maps are encoded to be used in GIS requires knowledge of cartography. Maps are not just artifacts; mapping is a process reflecting a way of thinking (Clifford and Valentine, 2003: 343).

2.4 Concepts of Participatory Approaches in IWRM

Based on the current scientific discussion on the concept of IWRM and the current water situation surrounding Namibia, there is a need for holistic participatory approaches to find the best practices and tools to support decision making process in the IWRM. The process involved empirical studies and conceptual work. The study entails both qualitative and quantitative approaches in order to cater for perception information, descriptive data as well as statistical and spatial data required for the project.

Participatory approaches are used, among other applications - to promote Community Based Natural Resources Management (CBNRM) on communal land. It uses information obtained from village mapping workshops to produce computer-generated maps (Taylor et al. 2006). In this project, the results of participatory approaches such as consultative meetings, FGDs and Workshops were appreciated. Emphases were also put on the development of GIS - based tool for IWRM for the Zambezi region.

Participation is promoted in order to encourage and reinforce local decision-making and local responsibilities in order to lead towards empowerment of local people, as it involves more equitable social redistribution to empower weak groups in access to, and control over resources, and to promote people's initiative, local control, and ownership of resources (McCall, 2004).

3. The Study Area

3.1 The Zambezi Basin (within Namibia)

The Zambezi basin was chosen because it is the fourth-largest river basin in Africa, after the Congo/Zaire, Nile and Niger basins (Food and Agriculture Organization, 1997). The size of the Zambezi catchment in Namibia measures about 20 000 km^2.

Zambezi region is surrounded by four other countries: Botswana to the south, Angola and Zambia to the north and Zimbabwe to the east as depicted in Figure 1. In border terms, the Zambezi region stretches 450 kilometres from west to east and ranges between 30 to 100 kilometres in width from north to south.

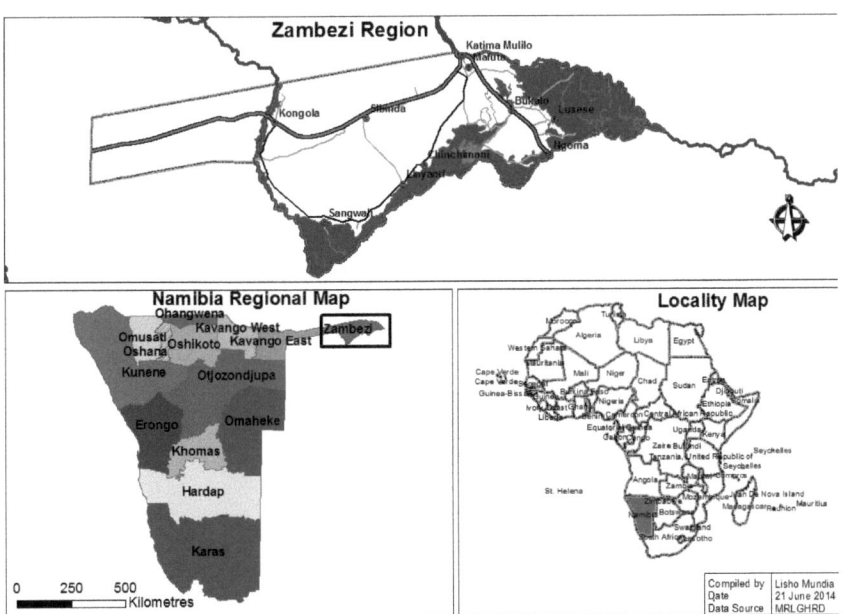

Figure 1: The study area map

3.2 Population Dynamics

In 2001, the Zambezi region had a population of 79.826 people, representing 4.4 percent of Namibia's total. The situation changed in 2013's housing and population survey as over 90,000 people live in the Zambezi region currently, representing about four percent of Namibia's population (Namibia Statistics Agency, 2013). The population consist mostly of subsistence farmers who make their living from the banks of the Zambezi, Kwando, Linyati and Chobe rivers.

The region's land surface area comprises 14 528 km², or 1.7 percent of the country. By 2001, the population density was still stable by 5.5 people per km², which, although higher than the national average of 2.1 people per km², is considerably lower than Ohangwena region (21.3 people per km²) and Oshana (18.7 people per km²), the regions with the highest population densities.

Population densities reach more than 100 people per km² around Katima Mulilo, but range between 5 and 50 people per km² along the trans-Caprivi highway from Kongola to Katima Mulilo and along the Katima Mulilo-Chinchimane-Sangwali-Kongola road. Similar densities can be observed around such settlements as Bukalo, Ngoma and Schuckmannsburg in the eastern parts of the region.

Large areas lying between the Trans-Caprivi highways and the Angolan border; the Trans-Caprivi highway and the Katima Mulilo-Sangwali-Kongola road have low population densities of less than one person per km² (Mendelsohn & Roberts, 1997). The main reason for these low population densities is that the latter areas have insufficient permanent water, and that the area north of the Trans-Caprivi highway is a proclaimed forest, restricting permanent settlement.

4. Interventions and Technical Approaches

4.1 Research Design

The research process involved conducting empirical studies. Since the study entailed both qualitative and quantitative approaches, a number of empirical research methods such as scientific literature studies, consultation meetings, FGD and GIS mapping were followed in order to unravel the interrelationship of the different water uses, availability, accessibility and water resource infrastructures. Fotheringham et al. (2000:4) states that, "quantitative geography consists of one or more of the following activities: the analysis of numerical spatial data; the development of spatial theory; and the construction and testing of mathematical models of spatial processes." The volume of high-quality data were considered for analyses in this project. The volume of high-quality data analysed in considerable depth and methodological precision were better than vast amounts of data superficially analysed (Bak, 2004).

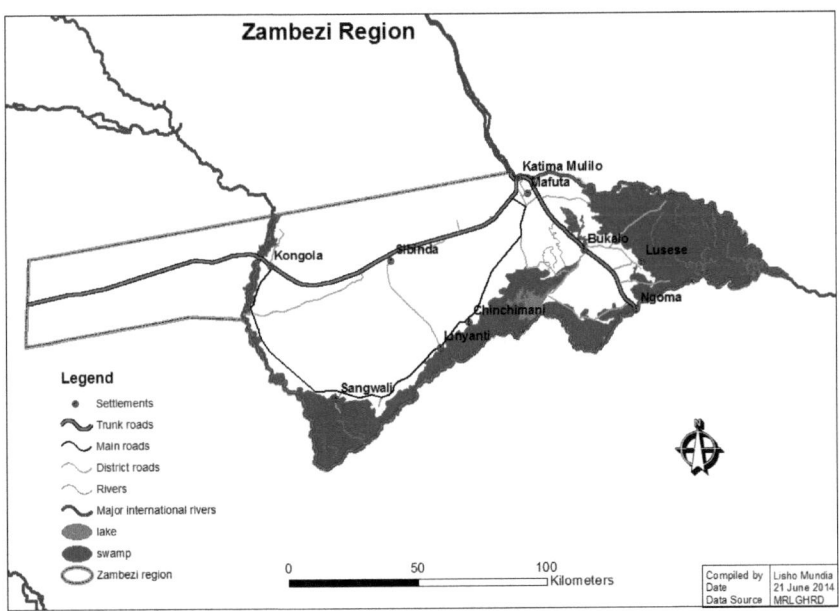

Figure 2: Base map of Zambezi region

11

The research included local participation and stakeholder involvement at different stages, such as data collection, verification and harmonisation of current and future scenarios. In this regard, natural water features in relation to human settlement were incorporated into the study (Figure 2). The stakeholder analysis was crucial for further participatory steps. These steps include the involvement of actual water users such as farmers and local communities for qualitative data gathering.

4.2 Population

The research was generalised in the worldwide context, with critical emphasis on Namibia's case in the theoretical framework. The practical aspect of the study, specifically the development of a GIS - based tool and application of the research methods was only done in Zambezi catchment area within Zambezi region. This is because the cross-countries legislation on data sharing and permits to carry out research of this nature in neighbouring countries could slow the research process and probably not make this research possible.

4.3 Sample

FGD used a minimum of five (5) attendants during its separate meetings. At least eight (8) FGD groups were held within Zambezi region, one FGD in each constituency. Local, regional and experts' knowledge were all helpful for this project for informed decision process and in understanding the process of IWRM. The study's output - "GIS - based tool" also have inputs from local, regional and national level. This was in order for the tool to be well measured and make its awareness to all stakeholders.

The targeted audience of different professions from different levels included representatives from local communities, authorities, NGOs and ministries. The participants were selected based on their specific knowledge relating to the research topic and questions. A designed FGD guideline with the support of open discussion questions was used. The following main aspects, as contained in FGD guideline questions were gathered and discussed during the FGD:

- Data on water resource, usage and infrastructures;
- Land access and rights;
- GIS - based tool in IWRM;
- Any other comments gathered from experts and local communities.

4.4 Research Instruments

Where there was no existing data, spatial data such as water points, telemetries points and other related water resources spatial data were collected using handheld Global Positioning System (GPS) instrument.

4.5 Procedure

The following research methods and processes in Table 2 were used to support GIS - based tool for proper knowledge gathering of the study.

Table 2: Data collection and processing procedures

Research Method (s)	Approach (es)	Outcomes and contribution to research objectives
Scientific literature review	Analyse the extent of a GIS - based tools and spatial data usage within IWRM implementation process	Knowledge about extent of usage, data availability usage within IWRM implementation process in Zambezi basin; identification of an appropriate tool framework for IWRM
Consultation meetings	Consultative meetings with different stakeholders involved in IWRM to understand local-to-regional decision processes through structured discussions	Knowledge and understanding about local-to-regional decision processes in IWRM
Focus Group Discussions (FGD)	The FGD used a minimum of five (5) attendants during its eight separate meetings within Zambezi basin in Namibia. The targeted audience of different professions included representatives from local,	Gathered local, farmers and experts' knowledge and first-hand information on IWRM, GIS and RS to support the decision making process in IWRM for improvement of water supply and availability

13

Research Method (s)	Approach (es)	Outcomes and contribution to research objectives
	regional and national levels	
Software Implementation of GIS - based tool; collection of necessary input datasets (RS, GIS-layers, statistics, tables, photo, etc.) including validation	Conceptualise, setup and implementation a user-friendly GIS - based tool using Geopublisher software. The current vector and raster data been updated by different ministries in Namibia were used	Production of a GIS - based tool help improve the integration of a variety of spatial information sources such as GIS and RS (i.e. thematic maps) as well as results of participatory techniques to help support the IWRM
GIS for the performance of spatial analysis within a GIS - based tool	Spatial analyses based on quantitative and qualitative data were performed, resulting in a variety of analytical outputs reflecting the current status of water resources with focus on water supply and availability in the Zambezi basin. Spatial analyses were done using commercial and open source software and tools (i.e. ArcGIS, Atlas Styler, Quatum GIS, etc). Results were incorporated into Geopublisher	The result of spatial analyses contributed to IWRM by enhancing measurement of current or required water infrastructure for the improvement of water supply and availability
Strengths, Weaknesses, Opportunities and	Evaluation of developed approaches, guidelines and tools	Strengths, Weaknesses, Opportunities and Threat (SWOT) analyses were

Research Method (s)	Approach (es)	Outcomes and contribution to research objectives
Threat (SWOT) analysis for the evaluation of the GIS - based support tool		conducted on the GIS - based tool in IWRM. The evaluation quantified the usage and functional capabilities perceptions of the GIS - based tool after presenting the sample to the participants, looking specifically at the understanding and implications of the tool

4.6 Data Needs and Requirements

Both spatial and non-spatial data are available in Namibia, however, it was imperative to critically choose which data is suitable for a GIS - based tool. This was because the tool requires complete, reliable, meaningful and updated data. The data about environment, water resources, socio-economic, climate, agriculture, vegetation, terrain model, water supply, tourism and other topographical data were the main datasets required for this project. The data were acquired from different sources, such as the Ministry of Agriculture, Water and Forestry, Ministry of Environment and Tourism, Ministry of Lands and Resettlements, Ministry of Regional, Local Government, Housing and Rural Development and the Namibia Statistical Agency.

4.7 Data Analysis

The spatial and non-spatial data were processed, managed and integrated using proprietary solutions (software). Among them included Quantum GIS, Microsoft office packages, ArcGIS 10.x software. The GIS - based tool was conceptualised, setup and implemented in a user-friendly environment using Geopublisher software. The software were chosen because they are suitable for processing, managing, maintaining, manipulating and visualising most IWRM data, compatible with most handheld GPS for downloading and uploading spatial data and are widely available.

In analysing the summaries of the FGD, it started by reading all the focus group discussions summaries in one sitting. This was then followed by looking for trends (comments that seem to appear repeatedly in the data) and surprises (unexpected comments that are worth noting). However, context and tone were considered as important in the reiteration of particular words. With regard to data on gender, age, constituency and professional level in water resource; these were gathered and documented using the demographic form. This helped measure the great details of respondents' demographic information. The analysis of this data was through the use of Microsoft Excel, where all categories were captured independently and analysed by constructing graphs and tables.

The production of the GIS - based support tool helped improve the integration of a variety of spatial information sources such as GIS and RS (i.e. thematic maps) as well as results of the FGD as descriptive themes to help support the IWRM. Geopublisher is an application to generate & publish multimedia atlases and has been developed to overcome the technical obstacles in publishing geo-data.

4.8 Implementation Plan and Schedule
Table 3 outline the brief activities, expected inputs, expected project outputs and the proposed timeline of each activity.

Table 3: Implementation plan and schedule

N o.	Activity	Input	Output	Timelin e
1.	Data collection (GIS data, Focus Group Discussions (FGDs)	Researcher and assistant	• Spatial data from NSA, MWF, MRLGHRD, MET and MLR; and • Descriptive data from both relevant organisations and local communities within Zambezi region	August 2014
2	Compiling a	Researcher and	Conceptual model	August

N o.	Activity	Input	Output	Timeline
.	conceptual model of an integrative GIS - based atlas	assistant	of the GIS – based digital atlas	2014
3 .	Literature survey and conceptual framework	Researcher and assistant	Information for incorporation into reports and scientific article for publications and dissemination to the public	September to November 2014
4 .	Collation and data (GIS and FGD) analysis	Researcher and assistant	Information for incorporation into the interactive GIS - based digital atlas	September 2014
5 .	Development of an interactive GIS - based digital atlas	• Researcher • Spatial and non-spatial data such as legislations and SWOT results	Interactive GIS - based digital atlas	October to January 2014
6 .	Technical training of about 15 staff members	Researcher and assistant	About 15 staff members in the Zambezi region received training on the GIS – based tool to be developed	November 2014
7 .	Compilation of the final project report	Researcher	About 50 pages technical report based on the project	February to March 2015
8 .	Compilation of the draft and final research article	Researcher and assistant	Scientific article for publication into the peer-reviewed journal	February to March 2015
9	Presentation of	Researcher	peer-reviewed	April

N o.	Activity	Input	Output	Timeline
.	reports/publication Processes		scientific article	2015

5. The Results Perspectives

5.1 GIS - Based Support Tool in Planning and Implementation of IWRM

De By, Georgiadou, Knippers, Kraak, Sun, Weir and van Westen (2004:165) stated that "large computerised collections of structured data are what we call a database." A database management system (DBMS) is a software package that allows the user to set up, use and maintain a database. Like a GIS allows setting up a GIS application, a DBMS offers generic functionality for database organisation and data handling (de By *et. al.*, 2004). The established GIS - based support tool provides custom, flexible and dedicated data management functionalities to decision-makers and in this particular case, people and institutions responsible for land use planning. The GIS - based support tool is designed for:

1. providing timely, transparent and easily readable outputs;
2. streamlining the production of maps, thus reducing time and cost requirements;
3. transforming data and information into knowledge.

The GIS - based support tool contains layers of graphic information and their relational databases that are transformed into maps. The information allows the user to view a composite of a specific area, adding an array of graphically oriented decision-making tools to the IWRM process. The interface of the GIS - based support tool is depicted in Figure 3. The extracted thematic map of the Zambezi region's water features is depicted in Figure 4.

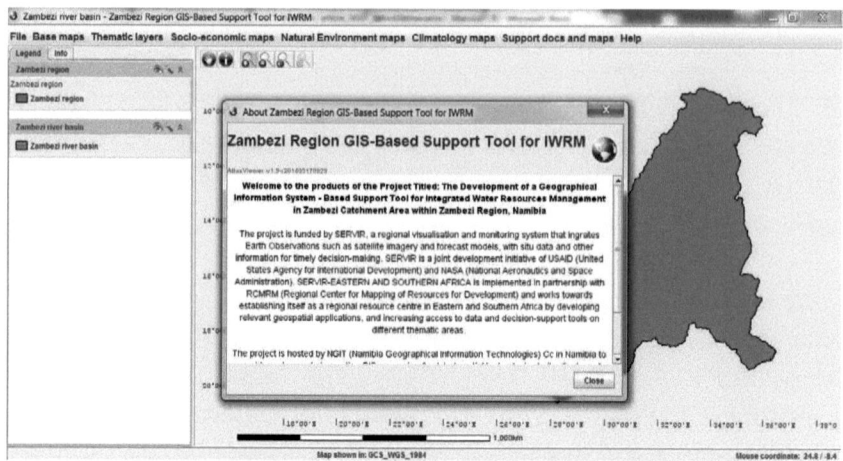

Figure 3: The interface of the GIS - based support tool for Zambezi region

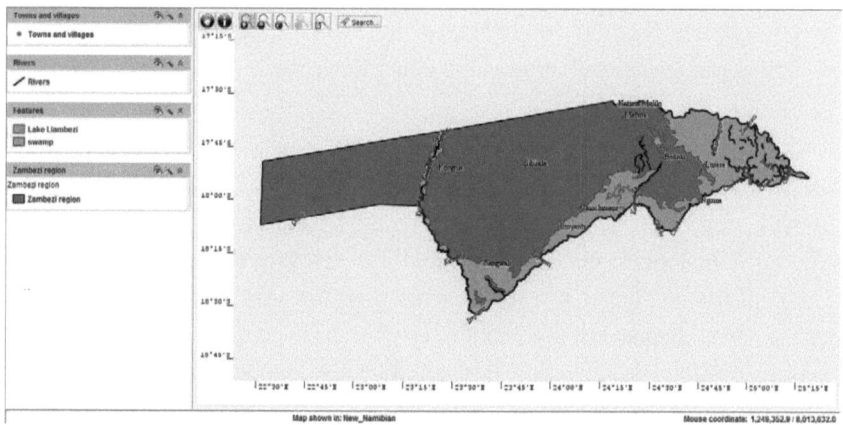

Figure 4: Screen shot of a thematic map of the water features of the Zambezi region created from the GIS - based support tool

A number of spatial layers, maps, legislation documents and reports were incorporated into the digital GIS – based support tool. The tool was designed to allow easy access to both spatial and non-spatial information concerning IWRM in the Zambezi region.

5.2 Strengths, Weaknesses, Opportunities and Threats (SWOT) Results

The SWOT analysis produced results from about 15 experts from different educational background, different organisations and knowledgeable individuals in the Katima Mulilo Urban Constituency. The results gathered are represented in Table 4.

Table 4: SWOT Analysis results from experts in the Zambezi region

Internal Factors	
Strengths	**Weaknesses**
<ul><li>Educated personnel</li><li>There is the availability and accessibility to spatial data and information.</li><li>Availability of computers resources in the region</li><li>Abundance of river water system</li><li>Good diverse natural resources</li><li>The region is good in agriculture, wildlife management, fishing and tourism</li><li>Committed people to venture into Geographical Information Sciences (GIS) and Integrated Water Resources Management (IWRM) in the region</li></ul>	<ul><li>Lack of GIS awareness, knowledge soft skills in the region</li><li>Lack of GIS software in the region</li><li>Lack of water coordination offices to take responsibilities and be accountable for water management</li><li>Human-wildlife conflict on water bodies</li><li>Lack of GIS and IWRM experts in the region</li><li>Lack of GIS and IWRM support from managements in the region</li></ul>
External Factors	
Opportunities	**Threats**
<ul><li>There is a young dynamic workforce willing to learn and explore GIS and IWRM programmes</li><li>GIS – based support tool brings the opportunity for experts to acquire new skills and collaboration with other organisations in Namibia</li><li>There are good Information Communication Technology (ICT)</li></ul>	<ul><li>Lack of finance to support GIS and IWRM programmes in the region</li><li>Poor accesses to information on programmes such as GIS and IWRM</li><li>Poor awareness programmes on GIS and IWRM.</li><li>Poor water infrastructures in some parts of the region</li><li>Emerging land disputes can</li></ul>

21

infrastructures in the region. • There is an availability of information on GIS and IWRM in the country • The availability of open source software is a great opportunity	hinder the programs of IWRM

5.3 Findings on Water Resources, Usage, Infrastructure and Land Tenure

The experts in the Katima Urban constituency stated that most parts of the Zambezi region have access to water for different purposes such as drinking, washing clothes, bathing, feeding animals and gardening. The water is accessed by different means such as groundwater boreholes systems, pipeline water system via taps (Figure 6), running water and open pits. It was stated by the experts in the Region that the longest distance some community members can walk to access water is 2.5 kilometres.

A major pipeline runs from Katima Mulilo to Kongola carrying water pumped from the Zambezi and Kwando rivers. Another pipeline running from Katima Mulilo to Linyanti is being completed, while the pipeline from Katima Mulilo to Ngoma is being constructed. The abundance of water sets the Zambezi region apart from the rest of the country. Out of the five permanent rivers in Namibia, three are in the Zambezi region: the Chobe, Kwando and Zambezi, as depicted in Figure 5).

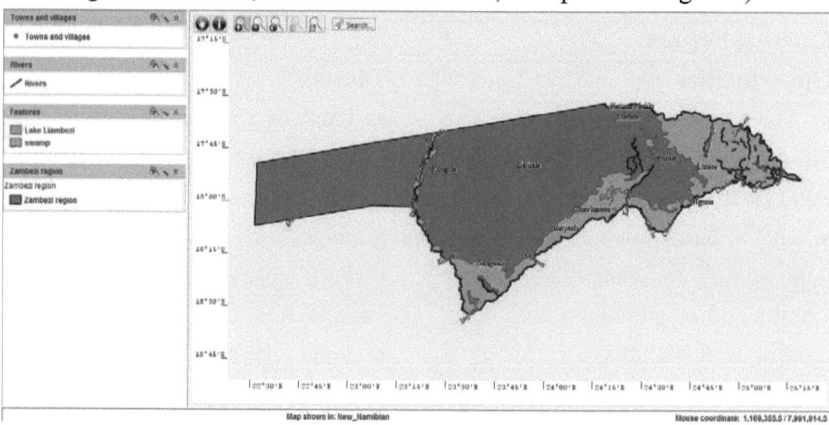

Figure 5: Permanently flowing rivers in the Zambezi region

Water quality is very variable throughout the region. The chemical composition of water rather than the yield potential is restricting the use of groundwater. It is frequently too salty, unpalatable or emanates from dirty wells and backwaters. Good quality water is generally found within 5 to 20 km from the rivers which recharge the aquifers. Recharge in the central parts of the region is low.

Figure 6: Water meters at Liselo Village in Katima Mulilo Rural Constituency

A number of non-functional water infrastructure were discovered with the Zambezi region. Among them, some boreholes and modern cattle drinking water canoes as depicted in Figure 7. Underground water is accessed using boreholes (see Figure 9) for human consumption and cattle in some parts of the Zambezi region.

Figure 7: None-functional water infrastructure at
Malengalenga in Judea Lyamboloma constituency

Some parts of the Zambezi region have poor water quality for human
consumption, these include areas like Sauzuo, Singalamwe, Lisikili, Nankuntwe,
Ikaba and Lizauli (Figure 8). The water in those areas is not purified. The Red Cross
of Namibia provides purification tablets to local community members who are using
direct water from rivers for human consumption. There are aging water infrastructure
in some parts of the Zambezi region contributing to poor water supply.

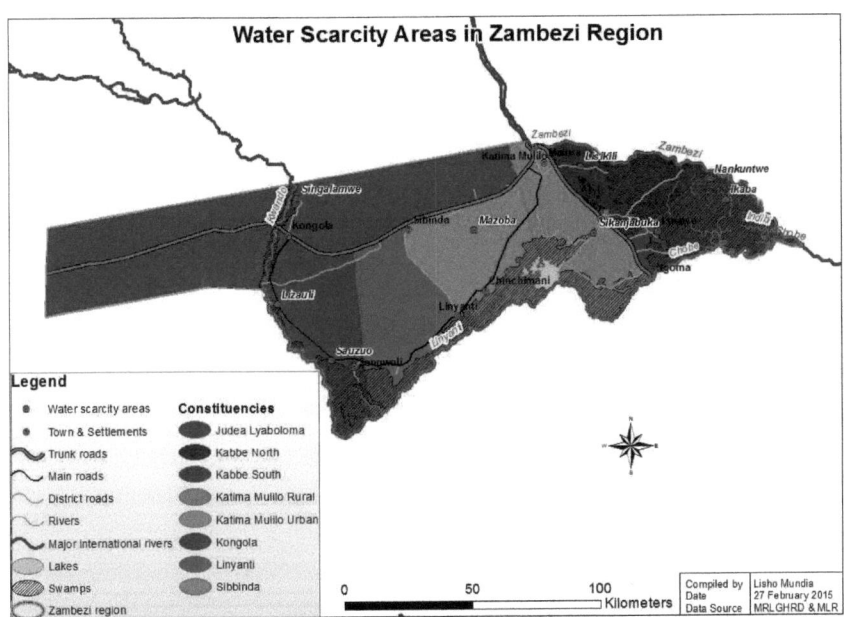

Figure 8: Water scarcity areas in the Zambezi region

Water is managed in tanks, reservoir and some earth dams at Lusese and Mazoba. These earth dams have now been dilapidated due to high density of livestock. In addition, wildlife in the surrounding areas has contributed to damaging the tanks, reservoirs and earth dams.

Figure 9: Borehole at Nsheshe village in Judea Lyamboloma constituency

Seasonal forest pans and depressions called swamps (Figure 10) play an important role in providing water to livestock and people during the rainy seasons and after rain seasons. In November when the rainy season starts, water collects in these pans and can remain there until August the following year. This allows livestock owners to graze their animals in wooded forest areas that have no developed groundwater sources. In addition, human beings use this water for drinking and washing clothes (Figure 10) in some parts of the Zambezi region.

Figure 10: Swamp water used for washing and drinking at Malengalenga

Land in the region is held on freehold in case of town land, customary land rights and leasehold land rights in case of communal land occupants. Various land uses exists in the region, such as residential, business, agriculture and nature reserves.

There is a challenge with supporting villages in communal land with water as some residents continue to establish villages in areas that are out of reach for possible installation of water infrastructures. The situation has resulted in lack of water as a lot of illegal land grabbing is taking place nearly every day.

A number of land disputes and conflicts have been registered with the Ministry of Lands and Resettlements and the Zambezi regional office in areas such as Sikanjabuka, Kuyugho (Ghau) and Kongola. The communal land boards in the Zambezi region resolve the land disputes and conflicts. Nevertheless, it has been a challenge for local communities to solve land disputes. This is because of the long period of time it takes for the communal land boards to resolve land disputes. The experts confirmed that it takes a maximum waiting period of between 5 to 6 months for land disputes and conflicts to be resolved in the region.

5.4 Livelihood and Agriculture Practices
Livelihoods in the Zambezi region are composed of several income streams. Farming, wages and salaries, cash remittances, pensions, tourism and fishing all

27

contribute towards sustaining the population. Significantly, only 21 percent of households reported that farming was their main source of income, whereas 30 percent indicated that farming was the main source for their salaries. Another 29 percent of households depend on business and non-farming activities, while only 15 percent and 6 percent of households respectively identified pensions and remittances as their main income source (Namibia Statistics Agency, 2013). The utilisation of natural resources such as fish and wildlife plays and important role, but has not been estimated separately in the population census. Migration does not appear to constitute a major source of income, as only 8 percent of the population were classified as short-term migrants. Looking at the remoteness of the region, it can be assumed that most of these emigrants are white-collar workers rather than itinerant workers.

Although about 70 percent of the population live in rural areas, less than 30 percent of the employed population are involved in subsistence agriculture (Namibia Statistics Agency, 2013).

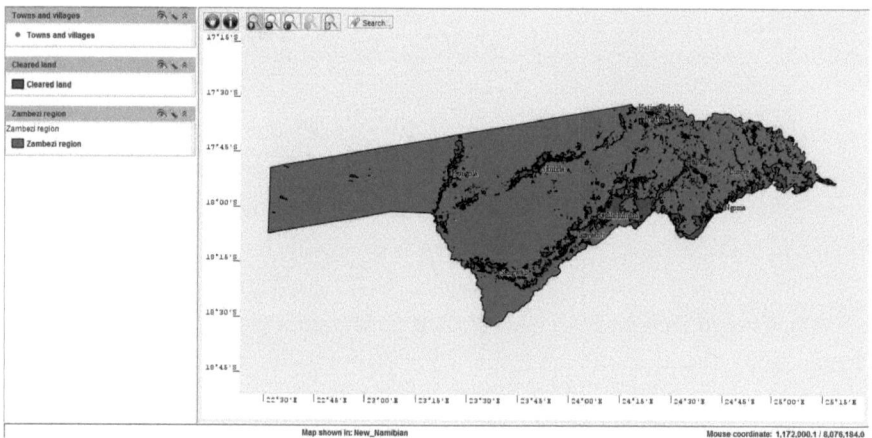

Figure 11: Land cleared for crop production in the Zambezi region

This makes the subsistence-farming sector an important provider of employment. Farming is dominated by rain fed cropping and livestock farming. Figure 11 shows the map of land cleared for crop production in the Zambezi region. The main crops grown in the Zambezi region are pearl millet (mahangu) and maize. During normal rainy seasons approximately 70 percent of the course gains produced consists of maize, while the remainder consists of mahangu and sorghum.

Cultivation in the Zambezi region is facing major constraints and threats. A major constraint is the poor quality of most soils: 70 percent of soils in the region have poor potential for cropping, while the best potential for cropping exists on only 15.6 percent of the soils. In the remaining areas, the potential for cropping is considered moderate (Mendelsohn & Roberts, 1997:16).

A lack of rainfall is the major threat to cultivation. Although the region has the highest long-term average annual rainfall in Namibia, ranging between 600 mm and 700 mm per year, precipitation varies dramatically between the rainy seasons, leading to regular droughts. In addition, rainfall is not always spread evenly across seasons. These risks are reflected in the variation of output of course grains, which have varied from a low of 1.400 mt in 1991/92 to a high of 16.000 mt during the 1999/2000 season. The likely severity of such seasonal shocks becomes clear when it is borne in mind that 27.300 ha were cultivated during the 1991/1992 season, the largest area planted during any one season since independence (Government of Namibia, 2001).

Total land planted each season as well as the size of fields cultivated per household also varies from year to year. Total area cultivated ranged between a high of 27.300 ha in 1991/92 to a low of 14.600 ha in 1997/98. It is reasonable to assume that these variations reflect farmers' assessments of the chances of a successful rainy season. Decreasing the land prepared for cultivation may be one strategy to minimise the negative impacts of a bad rainy season (Government of Namibia, 2001).

Crop farming in the Zambezi region is a 'low input - low output' system, reflecting the fact that investment in agriculture in terms of inputs is low, resulting in low yields and overall outputs. Improved seeds were used on approximately 30 percent of the area planted with mahangu, compared with 13 percent for maize and 4 percent for sorghum in the 1996/97 season. In contrast, traditional seeds were used on 49 percent of the area planted with mahangu, and 73 percent and 83 for maize and sorghum, respectively, during the same season. On the remaining areas, a mixture of traditional and improved seeds was used. Fertiliser and manure is applied in very few cases only (Government of Namibia, 2001). The use of improved seeds decreased substantially from the 1996/97 to the 2002/03 season (Government of Namibia,

2001). An estimated 10 percent of the area is prepared for cultivation by hand alone. In the 1996/97 season, 48 percent of the arable area was prepared by using oxen or donkeys alone, with 39 percent being prepared by comprising oxen/donkeys with hand preparation (Government of Namibia, 2001:41) Only 40 - 46 percent of households owned ploughs, while the number of households who owned oxen ranged between 37 percent and 52 percent between 1996/97 and 1998/99 seasons (Government of Namibia, 2001).

Altogether, the following factors: unreliable rainfall, moderate, soils, low inputs and low levels of agricultural technology - result in low yields of course grains in the Zambezi region. Maize yields varied between 30 kg per ha in the 1991/92 season to a high of 700 kg per ha in 1990/91. Corresponding yields for mahangu and sorghum were 70 kg (1991/92) and 460 kg (1999/00). Overall, the Zambezi region does not produce enough coarse grains to satisfy the consumption needs of the region. During the 2003/04 season, the cereal shortfall was estimated to be 7.300 mt (Government of Namibia, 2001).

In stark contrast to widely fluctuating yields and outputs of coarse grains, **livestock** numbers in the Zambezi region have shown a steady increase since 1990. Slightly fluctuations apart, the total number of cattle increased from 93.550 in 1990 to 149.030 in 2002. Figure 12 shows an estimated carrying capacity of the Zambezi region. About 10 to 15 hectares per large stock unit have limited supplies of water from rivers, giving pressures on pastures. Cattle owners are responding to this by taking their livestock into the forested areas where swamps fill up with water. During the rainy season in November, livestock is taken to these areas until the water dries up around July and August (Werner, Harris and Craig, 2002:17).

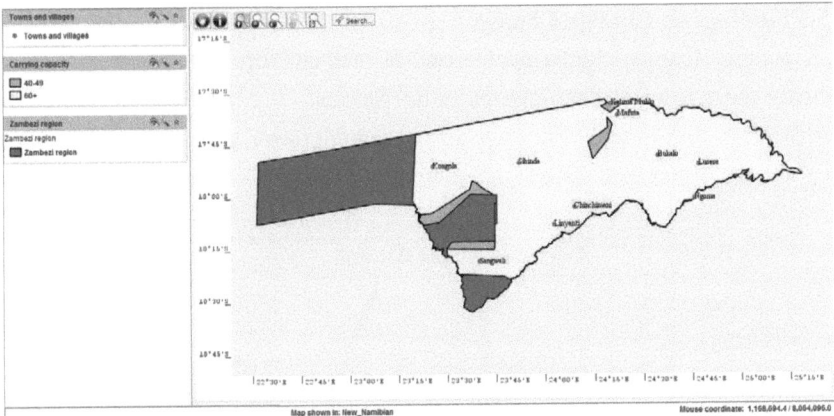

Figure 12: Livestock carrying capacity

Cattle are highly valued for a variety of reasons. To begin with, they are a source of cash income for most households. Until recently, the Meat Corporation of Namibia Ltd's (Meatco) abattoir in Katima Mulilo was the only formal market infrastructure in the Zambezi region. Since then an auction facility was constructed, which provides an alternative market to Meatco. In addition, informal butcheries serve as market for livestock.

Since the Zambezi region is classified as a veterinary control area, meat can only be exported to the rest of Namibia under strict veterinary supervision. In the year 2000, livestock was oversupplied (Government of Namibia, 2001:16-17).

Apart from livestock constituting a source of income for some people, cattle are extremely important in terms of crop production and providing draught animal power. Access to draught animal power enables farmers to cultivate large areas. Cattle also provide a measure of security for cash, consumption and cultural practices. They are often referred to as people's bank accounts, in that they serve as an investment and as a reserve in times of need, and people will invest in cattle when the opportunity arises. Cattle ownership is thus, a good indicator of livelihood security.

5.5 Findings on Climate Change

A country that is often characterised as hot and dry, the Zambezi region is distinctly more tropical than any region in Namibia. It enjoys a high rainfall with more than 500 millimetres of average annual rainfall (see Figure 13), less evaporation and a warmer winter than the rest of Namibia.

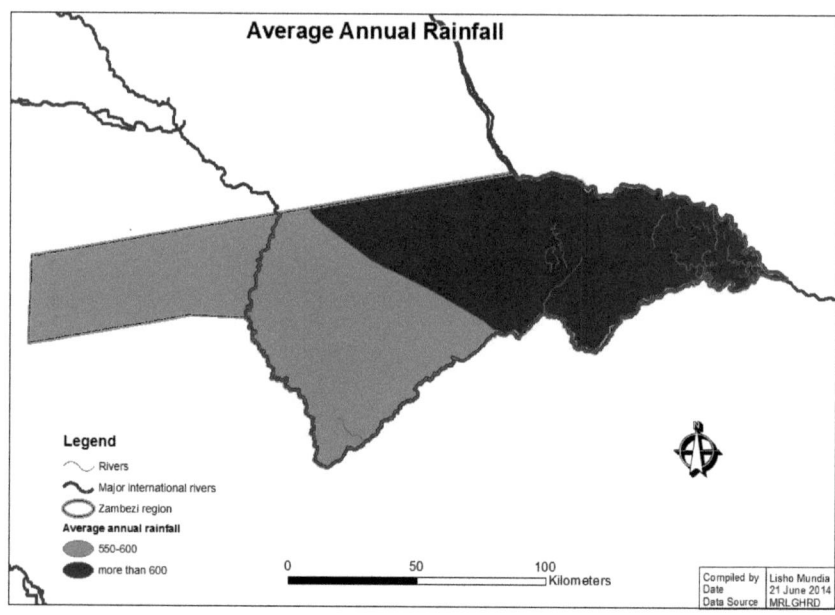

Figure 13: Average annual rainfall

Although the Zambezi region receives the highest rainfall in Namibia, it is still overwhelmed by rain that is highly variable from year to year and from one place to another. The region experience serious droughts from time to time. The Zambezi basin is considered to be vulnerable to climate variability.

The Zambezi region has a more tropical climate than any of the regions in Namibia. Summer temperatures peak in September, October and November, when they reach between 32°C and 35°C. As summer days are often cloudy, temperatures are fairly low, especially during the middle and last summer months. Average daily minimum temperatures vary between 20°C in summer and 5°C in winter (July). Clear skies in winter contribute to relatively high day temperatures. Frost is unusual, but does occur from time to time in low-laying areas (Mendelsohn & Roberts, 1997).

The Zambezi region receives its rain during the summer months. Small falls occur in September and October, while November usually provides sufficient rain for farmers to start ploughing. Rainfall peaks are between January and February, gradually decreasing towards April, when the rainy season comes to an end. Average rainfall in the Zambezi region ranges between 348 mm and 871 mm per year, and increases gradually from the south to north. Although this represents a high rainfall in Namibia's terms, its impact is reduced because rainfall is highly variable between and within seasons. Variability is not uniform across the region, and in the southernmost parts it is more pronounced.

The Zambezi region's **landscape** is characterised by open water, floodplains, riverine woodlands, mopane woodlands, Kalahari woodlands and impalila woodlands (Figure 14).

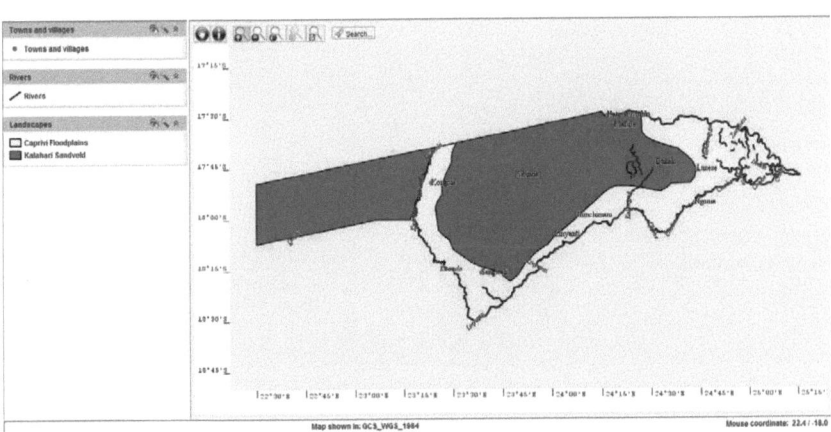

Figure 14: Landscapes of the Zambezi region

The Zambezi region is characterised by extreme flatness, falling from approximately 1 100 m in the west to 930 m near Impalila Island. Localised elevations in the form of dunes and dune valleys seldom exceed 30 m. The major features of the landscape are extreme Kalahari sands with their associated flood plains, channels and deposits. These have produced six major landscapes (Mendlelsohn & Roberts, 1997), namely areas of open water comprising the Kwando, Linyanti, Chobe and Zambezi rivers and their associated channels (see Figure 14).

Flood plains (see Figure 15), comprising of old river channels and low-lying areas close to the rivers. The riverine woodlands comprises of the riverine forests in the river valleys, whereas the mopane woodlands found in old river drainage channels are covered by wind-blown sand deposits. Kalahari woodlands cover, the extensive sandy plains of the eastern Zambezi region, as well as the western Zambezi region (commonly known as western Caprivi strip), which is dominated by small sand dunes and interdunes. The Impalila woodlands, which occur on the Impalila Island offer a unique scene to the eastern Zambezi region.

Figure 15: Caprivi region floodplains on the 8 May 2013

Soils vary from clay-loam in the flood plains (as depicted in Figure 16) through loam in the riverine woodlands, and from sandy loam in the mopane woodlands to sand in the Kalahari woodland areas.

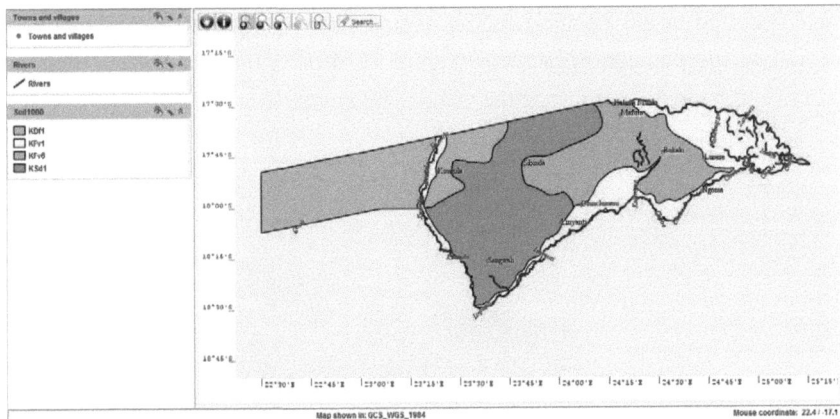

Figure 16: Soil types of the Zambezi region

Soils, flooding and fires influence vegetation in the Zambezi region (Mendelsohn & Roberts, 1997). Water drains easily through the sand, washing away nutrients. This leaves sands and grasses (Figure 17) low in nutrients. Areas that are flooded regularly hold water longer and often have a high content of organic materials. However, flooding restricts the growth of most woody vegetation, leaving extreme grasslands in the wettest areas, reeds and sedges predominate (Mendelsohn & Roberts, 1997).

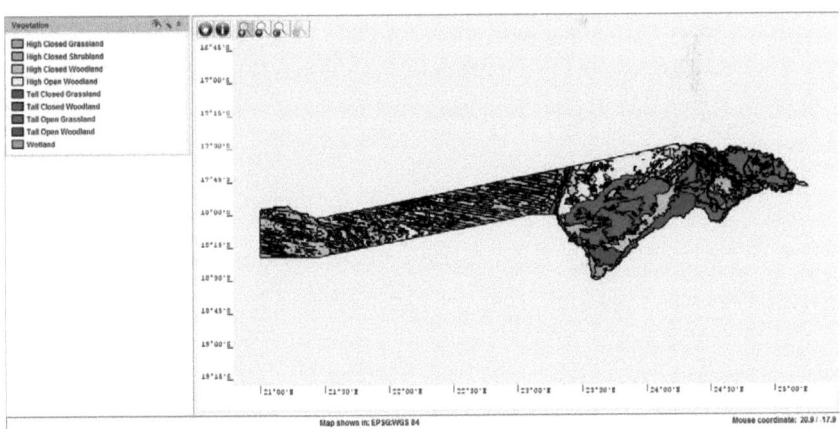

Figure 17: Vegetation of the Zambezi region

Many areas of dense woodland in the Zambezi region have been converted to open grasslands and shrub lands because of frequent fires. Although fires have always occurred in the Zambezi region, their frequency has increased dramatically, inter alia as the population increase. Figure 18 depicts the areas where fires have occurred and burned areas during the years 2000 and 2010.

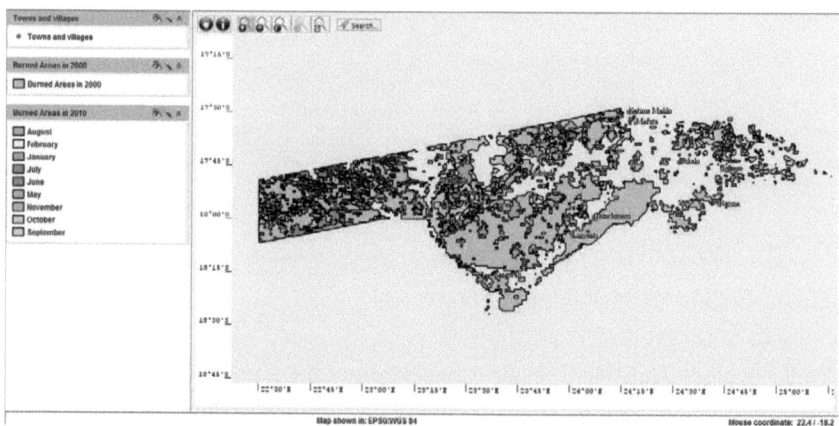

Figure 18: Burned areas in the Zambezi region

Fires have important benefits to people who depend on plants, but its frequency causes significant damage to vegetation. Young trees such as teak and kiaat for example, can no longer establish themselves. As a result, large areas in the region are now bush-encroached, leaving little space for grass growth (Mendelsohn & Roberts, 1997). This has also resulted in a number of flash floods (Figure 19) in some parts of the Zambezi region due to poor vegetated land for some water to be absorbed during rain seasons.

Figure 19: Flooded homestead in the village of Makolonga in the constituencies of Sibbinda on the 17 March 2014

The villages of Mpukano, Muzii and Nankuntwe in the eastern floodplains of Zambezi Region have always been affected with flood. The year 2014 was no exception for these villages as villages, schools, agriculture crop fields and grazing lands were flooded. This resulted in animals, people, and school going children being trapped from reaching their schools. In April 2014, schoolchildren have to use canoes to reach their schools (see Figure 20).

Figure 20: Flooded homesteads in the village of Nankuntwe – In the eastern floodplains of Zambezi Region on the 10 April 2014

Some parts of the Zambezi region experience long to short-term (flash) flood. These floods have some negative and positive impacts. The study results indicated that flood has the following negative and positive impacts with regard to flooding. Experts have indicated that the flood has the following negative impacts:

- Flooding some agricultural land for crop production and grazing.
- Flooding of homesteads (see Figure 21).
- It brings borne diseases such as malaria and cholera.
- Flood water breads dangerous poisons snakes.

Some positive impacts are:

- Good soil moisture after the flood.
- Fertilizes the land with good soil moisture.
- Floods bring water for animals as it creates more swamps.
- Floods bring more fish for the local community members when the water is subsiding.

There are challenges with regard to relocating the affected community members faced with flood. This is because there is scarcity of high elevated land in the region. Another challenge is that community members are not willing to relocate because there is scarce of fertile land for agricultural purpose. The community members refuse to relocate because they believe it is only temporary flood and prefer to stay in their ancestry land.

Figure 21: Rainwater (Flash flood) at Lusu village in Sibbinda Constituency – December 2014

In the eastern areas of the region, it is more predictable. Intra-seasonal variation refers to a situation where seasonal rainfall is spread unevenly across a season. This may impact negatively on crop production. Crops may get off to a good start but wither away in subsequent period of low rainfall. In the second place, evaporation is high. Some 2 500 mm of water evaporates in an average year in the Zambezi region, which is over four times the volume of water normally provided by rain. **Evaporation** rates increase during dry years (see Figure 22), when a lack of cloud cover contributes to high temperatures (Mendelsohn & Roberts, 1997). Despite some of these negative factors, the Zambezi region is ranked first in terms of agricultural potential. It has a dependable growing period of 120 days, and an average growing period of 135 days.

Figure 22: Drought in the Zambezi region, Sibinda constituency - 23 February 2015

6. Capacity Building and Alliances

6.1 Training of Technical Experts

About 15 staff members in the Zambezi region received training on the GIS – based tool. The tool was developed to contribute to IWRM process in the Zambezi region in terms of spatial data management, data processing and data representation.

The project team was devoted to decentralising knowledge to the local experts (Figure 23) in order to cater for the sustainability of the digital support tool for various applications in the region.

Figure 23: Training on the GIS - based support tool

6.2 Final Workshop Presentations

The final presentation on the project results were presented to about 15 staff members in the Zambezi region. The GIS – based support tool was also demonstrated to the participants on how to use it. The support tool was developed to contribute to IWRM process in the Zambezi region in terms of spatial data management, data processing and data representation. The tool was well received with great aspiration to use it in different applications such as land use, land administration, etc.

Figure 24: Some participants during the final workshop presentations

6.3 Alliance Opportunities

There is a need for alliances in the framework of IWRM project worldwide. The finding of this project indicates that there were possible organisations to form alliances with during the project, but the coordination to partner with other organisations within the region is cumbersome in terms of administration.

7. Recommendations and Conclusions

7.1 Recommendations

Based on the findings of this project, it can be concluded that not much research has been done on IWRM using participatory approaches and the use of GIS in Namibia. Therefore, there is a need for participatory approaches to be used in future IWRM project in order to cater for the inputs of local communities. GIS technology is vital for integration of different datasets. Throughout this research, multiple datasets were used and assembled. Most of these are spatial by nature and could be useful for other types of projects.

Participatory approaches such as consultative meetings, FGDs, workshops, GIS technology and the capacity to use these methods and technology have huge potential for more informed and more conscious decision-making. Nevertheless, if individuals or institutions are not empowered to make decisions then sustainable land water management cannot be well implemented. Establishing GIS – based support tool for IWRM that enable local people's involvement, enhances decisions are critical factors because sustainable water management must be built on stakeholder and water user involvement at local level.

7.2 Conclusions

In this project, participatory approaches such as consultative meetings, FGDs and workshops were identified as important towards IWRM in the region, particularly in sharing water resources ideas among participants and collecting data related to socio-economic issues. Participatory methods, if applied properly, allow the user to grasp the intangible and invisible through a concrete medium that can be shared with others. The developed GIS – based support tool is significant for all areas of IWRM and other natural resources management in the region. The study supports participatory approaches as potentially viable tools, techniques and methodology to clarify local communities' knowledge and create media that permit different voices to enter into dialogue with one another.

Another purpose of participatory approaches was to gather and share information about different negative and positive impacts of water resources among different participants in diverse areas of the Zambezi region. The participatory approaches methods proved to be an excellent process for allowing local people of all legitimate ages to engage with their surroundings and heritage. In an inspiring and motivating way, they were encouraged to use their water appropriately.

The developed GIS – based support tool activities offered other advantages of allowing experts from different organisation to learn about spatial data management, spatial data accessing, spatial data analysis and map-making. The GIS – based support tool also proved to be a catalyst in stimulating memory and in creating visible and tangible representations of the natural environment and land use. The time spent working with the experts allowed for greater clarity on meanings, and the relationship between natural water features and land use features. The tool is able to provide for data capturing of both land use and natural features.

7.3 Retrospections
The following retrospections are important for the success of IWRM in the Zambezi region:
- Engage local community members on water usage through participation
- Lack of visibility of the IWRM steering committee in the region
- There is a need to train technical staff members on IWRM concepts
- Develop skills to manage large-scale database system for IWRM in the region
- Engage line ministries, private sectors and local communities (water users) in the process of IWRM in the region
- Promotion of research funds to address technical challenges on water and wild-animal conflicts; flood and land use; and agricultural practices and climate change in the region
- Regional umbrella organisations to handle water infrastructure maintenance in some parts of the region, especially in Katima Mulilo
- Advertise for water conservation measures in the region

References

Bak, N. 2004. Completing your thesis: A practical guide. Cape Town: Van Schaik.

Biggs, D., Heyns, P., Klintenberg, P., Mazambani, C., Nantanga, K., & Seely M. 2008. Bridging Perspectives on IWRM in the Cuvelai Basin, 2008. on behalf of CuveWaters project and in corporation with Desert Research Foundation of Namibia (DRFN) & Institute for Social- Ecological Research (ISOE).

Clarke, K.C. 1997. Getting Started with Geographic Information Systems [Online]. Prentice Hall Companion Website. Santa Barbara: University of California. Retrieved on the 29th March 2010 from: http://wps.prenhall.com/esm_clarke_gsgis_4/7/1847/473050.cw/index.html

Clifford, N, J., and Valentine, G. 2003. Key Methods in Geography. SAGE Publications. London, Thousand Oaks, New Delhi.

De By, R.A., Georgiadou, P.Y., Knippers, R.A., Kraak, M.J., Sun, Y., Weir, M.J.C. and van Westen, C.J. 2004. Principles of geographic information systems : an introductory textbook. Enschede, ITC, 2004. ITC Educational Textbook Series 1, ISBN: 90-6164-226-4.

Food and Agriculture Organization. 1997. FAO's Information System on Water and Agriculture [Online]. Retrieved on 7 October, 2012 from: http://www.fao.org/nr/water/aquastat/countries_regions/namibia/index.stm

Fotheringham, A.S., Brunsdon, C. & Charlton, M. 2000. Quantitative geography: Perspectives on Spatial Data Analysis. London: SAGE Publications.

Global Water Partnership. 2008. A Handbook for Integrated Water Resources Management in Basins. Retrieved on the 22nd January 2011 from: www.gwpforum.org

Government of Namibia, 2001. Annual Agricultural Survey Report. Cental Bureau of Statistics 1998/99. Windhoek, Namibia.

IPCC. 2008. Climate change and water. IPCC Technical Paper VI. June 2008. Retrieved on the 29th January 2011 from: http://www.ipcc.ch/pdf/technical-papers/climate-change-water-en.pdf

Karnatak, H.C., Saran, S., Bhatia, K. & Roy, P.S. 2007. Multicriteria Spatial Decision Analysis in Web GIS Environment. Geoinformatica (2007) 11:407–429.

Kjelds, J., Jacobsen, T., Hughes, J. & Krejcik, J. 2005. Decision Support Tools For Integrated Water Resources Management. Basin Water Management. DHI Water & Environment, Agern Alle 5, 2970 Hørsholm, Denmark.

Klintenberg, P., Mazambani, C. and Nantanga, K. 2007. Integrated Water Resources Management in the Namibian Part of the Cuvelai Basin, Central Northern Namibia. CuveWaters Papers, No. 2 [Online]. Retrieved on the 6th July 2010 from: http://www.cuvewaters.net/ftp/cuve_2_klintenberg.pdf

Kluge, T., Liehr, S., Lux, A., Niemann, S., Brunner, K. 2006. IWRM in Northern Namibia – Cuvelai Delta. Final Report of a Preliminary Study [Online]. Retrieved on the 15th July 2010 from: http://www.cuvewaters.net/ftp/iwrm_final_report.pdf

McCall, M.K. 2004. Can Participatory-GIS Strengthen Local-level Spatial Planning? Suggestions for Better Practice [Online]. Retrieved on 6 May, 2011 from; http://www.gisdevelopment.net

McDonnell, R.A. 2008. Challenges for Integrated Water Resources Management: How Do We Provide the Knowledge to Support Truly Integrated Thinking? Water Resources Development, Vol. 24, No. 1, 131–143, March 2008. Taylor & Francis, Oxford OX1 3QY, UK.

Mendelsohn, J. and Roberts C. (1997). An environmental profile and atlas of Caprivi. Directorate of Environmental Affairs. Studio Scan, Pretoria, South Africa.

Namibia Statistics Agency. 2013. Namibia 2011 Population and Housing Census Basic Report. Windhoek, Republic of Namibia.

Robinson, A., Morrison, J.L., Muehrcke, P.C., Kimerling, A.J. and Guptill, S.C. 1995. *Elements of Cartography* (6th ed.). United States of America :John Wiley.

Tagg, J. and Taylor, J. 2006. PGIS and mapping for conservation in Namibia. Participatory learning and action, December 2006. Retrieved on the 22nd February 2011 from: http://pubs.iied.org/pdfs/G02931.pdf

Taylor, J, Murphy, C, Mayes, S, Mwilima, E, Nuulimba, N. and Slater-Jones, S. 2006. Land and natural resource mapping by San communities and NGOs: experiences from Namibia. Participatory learning and action, April 2006. Retrieved on the 22nd February 2011 from: http://www.iapad.org/publications/ppgis/PLA54_ch10_taylor_pp79-84.pdf

UN-Habitat. 2006. Meeting Development Goals in Small Urban Centres - Water and Sanitation in the World's Cities. 2006, United Nations Human Settlements Programme. London and Sterling,VA.

Wade, T, Sommer, S. 2006. A to Z GIS: An illustration dictionary of geographical information systems. (2nd ed.). California. ESRI Press.

Werner, D., Harris, T.M and Craig, W.J. 2002. Community Participation and Geographic Information Systems.

Annex: Workshop Programme, Training Programme and Participants

Annexure 1: Workshop Programme

Namibia Geographical Information Technologies Cc

Website: www.ngit.cc.na
TEL +264-(0)61-263160
E-mail: lisho@ngit.cc.na

P. O. Box 24250
Fax +264-(0)61-263169
Windhoek - Namibia

Quality Geomatics And GIS Services

WORKSHOP SCHEDULE

Topic: Geographical Information System - Based Support Tool for Integrated Water Resources Management in Zambezi Catchment Area within Zambezi Region, Namibia

Dates: August 29, 2014
Venue: National Youth Resource Centre, Katima Mulilo, Namibia

Project Brief

The Small Grants Programme of the SERVIR-Africa Project is an initiative under implementation at the Regional Centre for Mapping of Resources for Development (RCMRD) with support of United States Agency for International Development (USAID) and in partnership with National Aeronautics and Space Administration (NASA). The project is hosted by Namibia Geographical Information Technologies (NGIT) Cc in Namibia to provide and promote innovative GIS, surveying & related spatial technologies to its clients and business associates.

The objective of the Small Grants Programme is to improve environmental decision-making by dissemination of Earth observation data, products, and tools to empower governmental and government-affiliated institutions to make better-informed decisions. The anticipated outcome of these activities is to stimulate the innovative use of geospatial tools and information to translate science into sustainable policy and practice that addresses the environmental and developmental challenges posed by climate stresses.

The purpose of the workshop is to gather opinions, measure and share knowledge and experience with the participants on the concepts of Geographical Information System (GIS) - based tools for Integrated Water Resources Management (IWRM) in Zambezi region. The meeting will also create an opportunity to build networks amongst the participants as well as create forum towards mutual awareness as communities of practice.

Friday, August 29, 2014		
08:00 – 08:30	Arrival of Participants & Registration	
08:30 – 09:00	**OPENING SESSION** Welcome & Introductions (*Expectations, knowledge sharing & Outcomes*)	To be done by Lisho Mundia
09:00 – 09:45	Focus Group Discussions in groups (*water sources, usage, infrastructure, land rights, access, flood measure and impacts*)	Facilitator Lisho Mundia & Ivonne Makando
09:45 – 10:00	**GROUP PHOTO & TEA BREAK**	
10:00 – 10:30	Presentation of a conceptual framework of GIS – based tool for IWRM	To be presented by Lisho Mundia
10:30 – 11:15	SWOT Analysis in groups (*Share some Strengths, Weaknesses on GIS and IWRM i.e. skills, software, infrastructures, etc in your organisations; Some Opportunities and Threats outside your organisations*)	Facilitator Lisho Mundia & Ivonne Makando
11:15 – 11:45	Feedback on SWOT Analysis (*Presentations by groups*)	Facilitator Lisho Mundia
11:45 – 12:00	**CLOSING SESSION** Summary of the Workshop by Lisho Mundia, NGIT Next Step (meetings, consultation and short training) by Lisho Mundia, NGIT Closing Remarks by All	

CONTACTS, NGIT TEAM:
Mr. Lisho Mundia, Project Leader (lisho@ngit.cc.na)
Ms. Ivonne Makando, Project Assistant (ivonnemakando49@gmail.com)

Namibia Geographical Information Technologies Cc.
WEBSITE: www.ngit.cc.na
TEL +264-(0)61-263160
E-mail: lisho@ngit.cc.na

P. O. BOX 24250
FAX +264-(0)61-263169
WINDHOEK - NAMIBIA

Registration No. CC/2008/1542
Member: Lisho Chutuuh Mundia (Founder And Director)

Annexure 2: Training Programme

Namibia Geographical Information Technologies Cc

Website: www.ngit.cc.na	P. O. Box 24250
TEL +264-(0)61-263160	Fax +264-(0)61-263169
E-mail: lisho@ngit.cc.na	Windhoek - Namibia

Quality Geomatics And GIS Services

TRAINING SCHEDULE
Topic: Development of a Geographical Information System - Based Support Tool for Integrated Water Resources Management in Zambezi Catchment Area within Zambezi Region, Namibia

Dates: November 11 to 14, 2014
Venue: National Youth Resource Centre, Katima Mulilo, Namibia

Project Brief

The Small Grants Programme of the SERVIR-Africa Project is an initiative under implementation at the Regional Centre for Mapping of Resources for Development (RCMRD) with support of United States Agency for International Development (USAID) and in partnership with National Aeronautics and Space Administration (NASA). The project is hosted by Namibia Geographical Information Technologies (NGIT) Cc in Namibia to provide and promote innovative GIS, surveying & related spatial technologies to its clients and business associates.

The objective of the Small Grants Programme is to improve environmental decision-making by dissemination of Earth observation data, products, and tools to empower governmental and government-affiliated institutions to make better-informed decisions. The anticipated outcome of these activities is to stimulate the innovative use of geospatial tools and information to translate science into sustainable policy and practice that addresses the environmental and developmental challenges posed by climate stresses.

The purpose of the training is to provide technical training on how to manage, maintain, use, process and share the data of the GIS – based digital atlas on the concepts of Geographical Information System (GIS) - based tools for Integrated Water Resources Management (IWRM) in Zambezi region. The training is also a continuous effort in building a network amongst the participants as well as create forum towards water resources management and GIS technologies awareness in the Zambezi region.

Wednesday, November 12, 2014		
08:00 – 08:30	Arrival of Participants & Registration	
08:30 – 09:00	**OPENING SESSION** Welcome & Introductions (*Expectations, knowledge sharing & Outcomes*)	Facilitator: Lisho Mundia
09:00 – 09:30	**Presentation on the Concepts of GIS**	Facilitator: Lisho Mundia
09:30 – 10:00	**GROUP PHOTO & TEA BREAK**	
10:00 – 12:45	**Practical Session (Introducing GIS with QGIS)**	Facilitator: Lisho Mundia & Ivonne Makando
12:45 – 13:45	**LUNCH**	
13:45 – 15:45	**Practical Session (GIS Data Processing in QGIS Continues)**	Facilitator: Lisho Mundia & Ivonne Makando

Namibia Geographical Information Technologies Cc.
WEBSITE: www.ngit.cc.na   P. O. BOX 24250
TEL +264-(0)61-263160   FAX +264-(0)61-263169
E-mail: lisho@ngit.cc.na   WINDHOEK - NAMIBIA

Registration No. CC/2008/1842
Member: Lisho Christoh Mundia (Founder And Director)

Namibia Geographical Information Technologies Cc

Website: www.ngit.cc.na
TEL +264-(0)61-263160
E-mail: lisho@ngit.cc.na

P. O. Box 24250
Fax +264-(0)61-263169
Windhoek - Namibia

Quality Geomatics And GIS Services

Time	Activity	Facilitator
THURSDAY, NOVEMBER 13, 2014		
08:00 – 08:30	Arrival of Participants & Registration	
08:30 – 09:00	Welcome & Recap of day 1	Facilitator: Lisho Mundia
09:00 – 09:30	**Presentation on GIS Data and Analysis**	Facilitator: Lisho Mundia
09:30 – 10:00	TEA BREAK	
10:00 – 12:45	**Practical session (Analysis in QGIS)**	Facilitator: Lisho Mundia & Ivonne Makando
12:45 – 13:45	LUNCH	
13:45 – 15:45	**Practical session (Analysis and Mapping in QGIS)**	Facilitator: Lisho Mundia & Ivonne Makando
FRIDAY, NOVEMBER 14, 2014		
08:00 – 08:30	Arrival of Participants & Registration	
08:30 – 09:00	Presentation session (Concepts of the GIS – based support tool)	Facilitator: Lisho Mundia
09:00 – 09:30	**Practical session (Using the GIS – based support tool)**	Facilitator: Lisho Mundia
09:30 – 10:00	TEA BREAK	
10:00 – 12:00	**Practical session (Using the GIS – based support tool continuous)**	Facilitator: Lisho Mundia & Ivonne Makando
12:00 – 12:15	**CLOSING SESSION** Summary of the training by Lisho Mundia Next Step (meetings, workshop and official delivery of the digital atlas) by Lisho Mundia Closing Remarks by All	

CONTACTS, NGIT TEAM:
Mr. Lisho Mundia, Project Leader (lisho@ngit.cc.na)
Ms. Ivonne Makando, Project Assistant (ivonnemakando49@gmail.com)

Annexure 3: Final Workshop Presentations Schedule

Namibia Geographical Information Technologies Cc

Website: www.ngit.cc.na | P. O. Box 24250
TEL +264-(0)61-263160 | Fax +264-(0)61-263169
E-mail: info@ngit.cc.na | Windhoek - Namibia

Quality Geomatics And GIS Services

WORKSHOP SCHEDULE

Topic: Development of a Geographical Information System - Based Support Tool for Integrated Water Resources Management in Zambezi Catchment Area within Zambezi Region, Namibia

Dates: April 7, 2015
Venue: National Youth Resource Centre, Katima Mulilo, Namibia

Project Brief

The Small Grants Programme of the SERVIR-Africa Project is an initiative under implementation at the Regional Centre for Mapping of Resources for Development (RCMRD) with support of United States Agency for International Development (USAID) and in partnership with National Aeronautics and Space Administration (NASA). The project is hosted by Namibia Geographical Information Technologies (NGIT) Cc in Namibia to provide and promote innovative GIS, surveying & related spatial technologies to its clients and business associates.

The objective of the Small Grants Programme is to improve environmental decision-making by dissemination of Earth observation data, products, and tools to empower governmental and government-affiliated institutions to make better-informed decisions. The anticipated outcome of these activities is to stimulate the innovative use of geospatial tools and information to translate science into sustainable policy and practice that addresses the environmental and developmental challenges posed by climate stresses.

The purpose of the workshop is to share learned lessons, knowledge and experience with the participants on the concepts of Geographical Information System (GIS) - based tools for Integrated Water Resources Management (IWRM) in Zambezi region. The workshop will also strengthen networks amongst the participants as well as create forum towards mutual awareness as communities of practice.

Tuesday, April 07, 2015		
08:00 – 08:30	Arrival of Participants & Registration	
08:30 – 09:45	**OPENING SESSION** Welcome & Introductions (*About the project*)	To be done by Lisho Mundia
09:45 – 10:00	**GROUP PHOTO & TEA BREAK**	
10:00 – 10:30	Presentation of Project Findings	To be presented by Lisho Mundia
10:30 – 11:15	Discussions and Questions	Facilitator Lisho Mundia & Ivonne Makando
11:15 – 11:45	1. Presentation of the GIS – based tool for IWRM 2. Introducing other online hydrological products in Africa	Facilitator Lisho Mundia
11:45 – 12:00	**CLOSING SESSION** Summary of the Workshop by Lisho Mundia, NGIT Next Step (deliverable means for the digital atlas & project reports) by Lisho Mundia, NGIT Closing Remarks by All	
12:00 – 13:00	**LUNCH**	

CONTACTS, NGIT TEAM:
Mr. Lisho Mundia, Project Leader (lisho@ngit.cc.na)
Ms. Ivonne Makando, Project Assistant (ivonnemakando49@gmail.com)

Namibia Geographical Information Technologies Cc.
WEBSITE: www.ngit.cc.na | P. O. BOX 24250
TEL +264-(0)61-263160 | FAX +264-(0)61-263169
E-mail: info@ngit.cc.na | WINDHOEK - NAMIBIA

Registration No. CC/2008/1842
Member: Lisho Christoh Mundia (Founder And Director)

iv

Annexure 4: Workshop, SWOT Analysis and Training Participants

No.	Name	Gender (F/M)	Position	Organisation	Email
1.	Malena Musialela	M	IT Technician	KMTC	malena@kmtc.org.na
2.	Syson Ntema	M	Assistant Administrator: Land registration	MLR Katima Mulilo	sntema@gmail.com
3.	Benitha Kwala	F	GIS Technician	MLR Katima Mulilo	benitha@live.com
4.	Benigina Kafe	F	Assistant Town Planning Officer	KMTC	ipbkafe@gmail.com
5.	Hendrix Minyoi	M	Assistant Administrator: Land registration	MLR Katima Mulilo	hminyoi@dudumail.com
6.	Nestor Sibeya	M	Chief Education Officer	Ministry of Education	msibeya@yahoo.com
7.	Avens Kabula	M	Assistant Administrator: Land registration	MLR Katima Mulilo	Akabula61@gmail.com
8.	Eiono Shanika	M	GIS Analyst	NSA	autyshaanika@gmail.com
9.	Cletius Mubita	M	Chief Development Officer	Zambezi Regional Officer	cletiusmubita@yahoo.com
10.	Bevery Saushini	F	Chief Administration Officer	MLR Katima Mulilo	mubu.saushini@gmail.com
11.	Daniel Mbala	M	Chief Development Officer	Zambezi Regional Officer	
12.	John Kambimbi	M	Chief Works Inspector	Ministry of Education	johnkambimbi@gmail.com
13.	Joseph Mulisa	M	Water Superintendent	Namibia Water Cooperation Ltd	MulisaJ@namwater.com.na

Annexure 5: Conceptual Design for the GIS – Based Digital Atlas

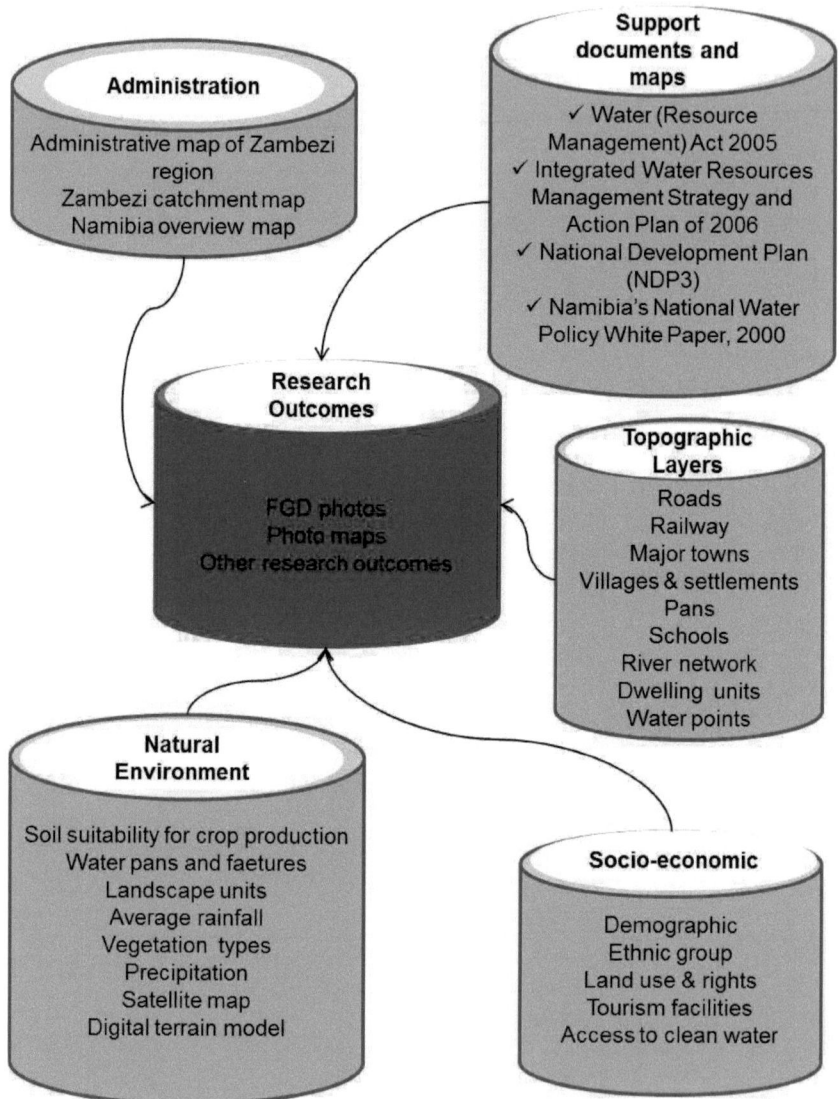

Printed by Books on Demand GmbH, Norderstedt / Germany